Pupil Book 3B

Series Editor: Peter Clarke

Authors: Elizabeth Jurgensen, Jeanette Mumford, Sandra Roberts

Contents

3-digit numbers

Recognise the place value of each digit in a 3-digit number

 Challenge 1

1 Write the numbers shown by the Base 10.

You will need:
• Base 10 blocks

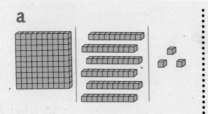
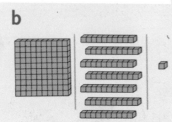
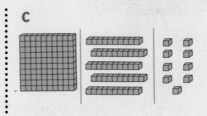

a   b   c

2 Use your own Base 10 to make the numbers below.

| a | 126 | b | 164 | c | 119 | d | 237 | e | 242 | f | 266 |

 Challenge 2

1 Write the numbers shown by the Base 10.

a   b   c   d

2 Write the numbers that have been separated into 100s, 10s and 1s.

| a | 200 | 10 | 9 | b | 400 | 50 | 5 | c | 500 | 70 | 9 |
| d | 600 | 20 | 9 | e | 700 | 70 | 3 | f | 800 | 0 | 8 |

 Challenge 3

1 Write the numbers that have been separated into 100s, 10s and 1s.

| a | 80 | 500 | 3 | b | 4 | 800 | 30 | c | 10 | 500 | 1 |
| d | 900 | 9 | 30 | e | 70 | 800 | 4 | f | 80 | 700 | 6 |

2 I'm thinking of a number. The 100s digit is between 4 and 6, the 10s digit is 7 and the 1s digit is an odd number. Write as many possible answers as you can.

Ordering numbers to 1000

Compare and order numbers up to 1000

Challenge 1

Order each set of numbers, smallest to largest.

a 254, 168, 517, 429   b 572, 266, 103, 381   c 467, 236, 423, 351

d 109, 541, 377, 218   e 276, 317, 180, 441   f 288, 371, 125, 452

g 151, 132, 169, 124   h 183, 137, 142, 193   i 258, 231, 227, 263

j 238, 254, 286, 207   k 283, 253, 213, 243   l 376, 384, 362, 390

Challenge 2

1 Order each set of numbers, smallest to largest.

a 265, 832, 163, 589, 322   b 831, 538, 529, 187, 276

c 945, 276, 254, 723, 433   d 376, 512, 352, 589, 312

e 478, 492, 382, 319, 408   f 821, 903, 962, 855, 965

g 444, 402, 414, 490, 476   h 899, 843, 856, 809, 819

2 For each number, write a 3-digit number that is larger and one that is smaller.

a 456   b 378   c 245   d 890

e 799   f 802   g 921   h 943

Challenge 3

1 Write a set of instructions for ordering 3-digit numbers.

2 Using the cards, make two 3-digit numbers that are
   larger than the number on each of the whiteboards above.

5

Using money to show 3-digit numbers

Represent and estimate numbers using money

Challenge 1

What amounts are shown here? Write each answer as pence.

a   10p 10p 10p 1p 1p 1p 1p

b   10p 10p 10p 10p 1p 1p 1p

c   10p 10p 10p 10p 10p 1p

d   10p 10p 1p 1p 1p 1p 1p

e   10p 10p 10p 10p 1p 1p 1p 1p 1p 1p 1p

f   £1 10p 10p 1p 1p 1p   g   £1 10p 10p 10p 1p 1p 1p

h   £1 £1 10p 10p 10p 1p   i   £1 £1 10p 10p 1p 1p 1p 1p 1p 1p

> **Example**
>
> £1 + 10p + 10p + 10p + 10p + 1p = 141p

Challenge 2

1   What amounts are shown here? Write each answer as pence.

a   £1 £1 10p 10p 10p 1p 1p 1p

b   £1 £1 10p 10p 10p 10p 1p 1p 1p 1p 1p

c   £1 £1 £1 10p 10p 1p 1p 1p 1p

d   £1 £1 £1 £1 10p 10p 10p 10p 1p 1p

e   £1 £1 £1 £1 £1 10p 10p 10p 10p 10p

f   £1 £1 £1 £1 £1 £1 10p 1p 1p 1p

> **Example**
>
> £1 + £1 + 10p + 10p + 1p = 221p

2   What coins would you need to make these amounts?

a   245p   b   316p   c   453p   d   507p   e   732p   f   663p

Challenge 3

Imagine you had 8 coins in your pocket –
they could be £1 coins, 10p coins or 1p coins.
What amount of money might you have?
Find different amounts.

> **Example**
>
> £1 + £1 + £1
> + 10p + 10p + 10p
> + 1p + 1p = 332p

Get the order

Compare and order numbers up to 1000

Work in threes.

- One player secretly writes a number on each of the pieces of paper.
- The other two players write the numbers 1–6 as a list in their books. At the top of the page, write: 'largest' and at the bottom write: 'smallest'.
- The player with the numbers reads them out one at a time.
- The other players decide where to write the number on their list.
- Players keep going until all the numbers have been read out.
- If a number cannot be put in order, write it next to the list.

Challenge 1

1 Secretly write a 2-digit number on each of the six pieces of paper.

2 How many of the numbers are in order?

3 Play again, this time with a different player writing the six secret numbers.

You will need:
- six small pieces of paper

Challenge 2

1 Secretly write a 3-digit number on each of the six pieces of paper.

2 Explain how you made your choices about where to write the numbers.

3 How can you improve your ordering next time?

You will need:
- six small pieces of paper

Challenge 3

1 Secretly write a 3-digit number on each of the ten pieces of paper. The other players write the numbers 1-10 as a list.

2 Which numbers did you find hardest to order? Explain why.

3 Write three top tips for playing this game. Try them out in the next round.

You will need:
- ten small pieces of paper

Café totals

Add amounts of money

How much would it cost to buy these items from the café?

Challenge 1

a apple juice and fairy cake

b orange juice and banana

c sausage roll and apple juice

d cheese sandwich and banana

e orange juice and fairy cake

f sausage roll and banana

apple juice 34p

orange juice 37p

cheese sandwich 58p

sausage roll 45p

fairy cake 42p

banana 25p

Challenge 2

a cup of tea and tuna sandwich

b ice cream and cheese toastie

c apple pie and cup of tea

d tuna sandwich and apple pie

e ice cream and cheese toastie

f apple pie and ice cream

apple pie 50p

cheese toastie 80p

ice cream 72p

tuna sandwich 74p

cup of tea 65p

Challenge 3

1 a egg sandwich and mango juice

b cheese toastie and milkshake

c hot chocolate and egg sandwich

d carrot cake and mango juice

e egg sandwich and cheese toastie

f milkshake and carrot cake

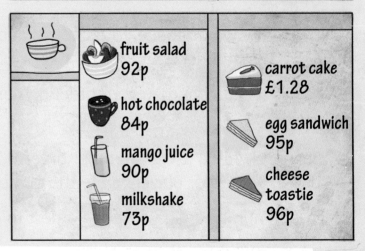

fruit salad 92p

hot chocolate 84p

mango juice 90p

milkshake 73p

carrot cake £1.28

egg sandwich 95p

cheese toastie 96p

2 I bought three things in the café. I spent a total of £2.96. What did I buy?

Café change

Subtract amounts of money to give change

mini pizza 84p	carrot cake £1.28	apple juice 34p	fairy cake 42p
ice cream 72p	cheese toastie 96p	orange juice 37p	banana 25p
cup of tea 65p		tuna sandwich 74p	fruit salad 92p
milkshake 73p	slice of toast 21p	sausage roll 45p	hot chocolate 84p

Challenge 1

Daniel has 50p. How much change will he get if he buys these things at the café?

a   apple juice   b   banana   c   fairy cake

d   slice of toast   e   sausage roll   f   orange juice

Challenge 2

1   Youssef has £1 (100p). How much change will he get if he buys these things?

a   fairy cake   b   cup of tea   c   mini pizza

d   ice cream   e   tuna sandwich   f   apple juice

2   Explain how to find change.

Challenge 3

1   Irene has £2 (200p). How much change will she get if she buys these things?

a   fruit salad   b   orange juice   c   cheese toastie

d   hot chocolate   e   carrot cake   f   milkshake

2   How much change will Sam get from £2 (200p) if she buys these things?

a   orange juice and carrot cake   b   hot chocolate and cheese toastie

c   milkshake and fruit salad

Buying fruit

Add and subtract amounts of money

apple 42p

orange 37p

plums 92p per 100 g

grapes 75p per 100 g

bananas 68p per 100 g

pear 29p

melon 146p

strawberries 105p per 100 g

Challenge 1

You have £1 (100p). How much change will you get if you buy this fruit?

a 1 orange	b 100 g of grapes	c 1 apple
d 100 g of plums	e 1 apple and 1 orange	f 2 pears
g 1 pear and 1 orange	h 2 apples	i 2 oranges

Challenge 2

You have £2 (200p). How much change will you get if you buy this fruit?

a 1 melon	b 100 g of strawberries
c 1 orange and 1 apple	d 200 g of grapes
e 100 g of plums and a pear	f 200 g of bananas
g 1 melon and 1 orange	h 100 g of strawberries and 100 g of grapes

Challenge 3

You have £5 (500p). How much change will you get if you buy this fruit?

a 200 g of strawberries	b 1 melon and 1 orange
c 100 g of strawberries and 100 g of plums	d 2 melons
e 1 apple and 100 g of bananas	f 200 g of plums
g 3 apples	h 300 g of grapes

Furniture shopping

- Add and subtract amounts of money
- Solve word problems involving money and reason mathematically

Mina, Lily and Jacob are buying new items for their bedrooms. Mina has £100, Jacob has £200 and Lily has £500 to spend.

Work out these money problems. Show your working out.

Challenge 1

a Jacob buys a rug and a mirror. How much does he spend?

b Lily chooses the most expensive item in the shop. How much does it cost?

c Mina buys a desk. How much change will she get?

d Lily buys a desk and a rug. How much do they cost?

Challenge 2

a Mina wants a bookcase and a mirror. How much more money does she need?

b Lily buys the two most expensive items. What will the total be?

c Jacob chooses the cheapest item in the shop. How much change will he get?

d Lily buys a rug, mirror and a desk. How much will she spend?

Challenge 3

a Lily buys a bed and a mirror. How much change does she get?

b Mina wants a bookcase and a wardrobe. How much more money does she need?

c Jacob buys two rugs and a mirror. How much change will he get?

d If Mina and Jacob put their money together, can they buy 3 armchairs?

Drawing and naming shapes

Draw and name 2-D shapes

You will need:
- Resource 21:
  3 x 3 pinboards
- ruler

Challenge 1

1 Name these shapes.

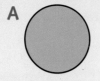

A   B

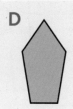
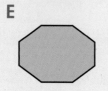

C   D   E   F   G   H

2 Draw these shapes on the 3 x 3 pinboard.

  a triangle **b** square **c** rectangle **d** pentagon **e** hexagon

Challenge 2

1 Name these shapes.

You will need:
- ruler
- squared paper

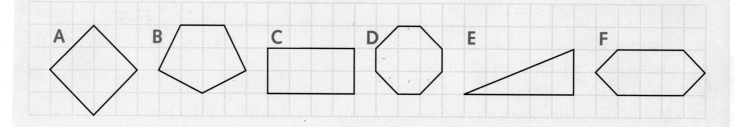

A   B   C   D   E   F

2 Draw these shapes on squared paper.

  a a square with sides of 4 cm   **b** a rectangle with sides of 3 cm and 4 cm
  c 2 different pentagons   **d** 2 different hexagons

Challenge 3

Use the 3 x 3 pinboard to draw ten different 4-sided shapes.

You will need:
- Resource 21: 3 x 3 pinboards
- ruler

Matching 2-D shapes

Make shapes that match a property

You will need:
- Resource 22:
  Matching shapes (1)
- scissors
- squared paper
- ruler
- blue and red pencils

1 Using Resource 22:
   Matching shapes (1), join
   two triangles along matching
   edges to make these shapes.

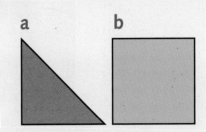

a   b

2 a Draw the two shapes above on squared paper.

   b Rule a blue line to show how the triangles fit together.

   c Circle the right angles in red.

Work with a partner.

a Cut out the rectangle and two triangles on
   Resource 24: Matching shapes (2).

b Make the shapes below using the rectangle
   and the right-angled triangles.

c Draw each shape showing how you made it
   using the rectangle and triangles.

You will need:
- Resource 24:
  Matching shapes (2)
- scissors
- squared paper
- ruler

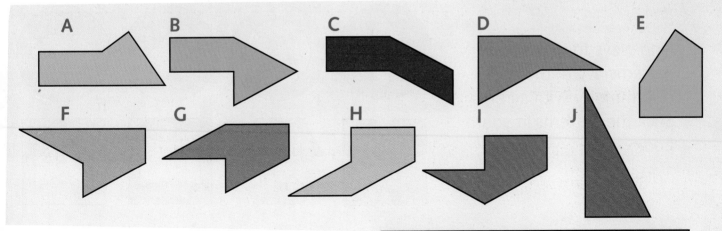

A   B   C   D   E

F   G   H   I   J

Look at the ten shapes in Challenge 2.
Copy and complete the table.

Property	Shape
has one right angle	
has more than one right angle	A,

Paper shapes

Make shapes using folding and cutting

You will need:
- paper squares or circles
- scissors
- ruler
- pen
- glue

Challenge 1

1 Fold the paper square into quarters.

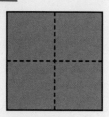

2 Make these patterns by folding and cutting. Mark the lines of symmetry with a pen. Stick the patterns in your book.

a   b   c

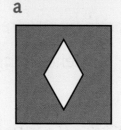  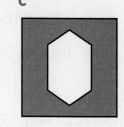

Challenge 2

1 Fold the paper circle and squares into quarters.

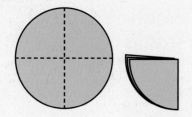

2 Find ways to make these patterns. Mark the lines of symmetry with a pen. Stick the patterns in your book.

a   b   c   d

   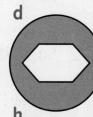

e   f   g   h

   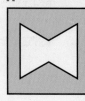

Challenge 3

Design two more patterns of your own. Mark any lines of symmetry and stick them in your exercise book.

Properties of 2-D shapes

Describe the properties of 2-D shapes

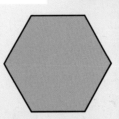
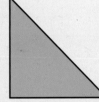

Challenge 1

Read the clues. Find the shape. Write its name.

- I have 5 equal sides.
- I have 1 right angle and 3 sides.
- I have equal sides and 4 right angles.
- I have 6 equal angles all greater than a right angle.

Challenge 2

1 Take two triangles and one regular shape from Resource 61: 2-D shapes. Fit them together to make these shapes. Name each shape.

You will need:
- Resource 61: 2-D shapes
- right-angle tester
- ruler
- scissors

A   B   C   D   E

2 Solve these puzzles about the shapes in Question 1.

a I have 2 angles greater than a right angle and 2 angles less than a right angle. Which shape am I?

b I have 1 right angle and my 6 sides are equal in length. Which shape am I?

Challenge 3

Andy said, "I can use 12 rods to make 5 squares." Is he correct? Sketch the squares you make.

You will need:
- Twelve 10 rods from Base 10 materials

15

Counting in steps of 2, 4 and 8

Count in multiples of 2, 4 and 8

Challenge 1

Write the missing numbers.

a   2, 4, _____ , 8, _____ , _____ , _____ , _____ , 18, _____ , _____ , 24

b   4, _____ , _____ , 16, _____ , _____ , _____ , _____ , 36, _____ , 44, _____

c   8, _____ , 24, _____ , _____ , _____ , 56, _____ , _____ , _____ , 88, _____

Challenge 2

1   Find the multiples of each number and write them in order, smallest to largest.

a multiples of 2   b multiples of 4   c multiples of 8

21 2 18   32 12 36   8 72 48
6 4 14   4 34 16   36 24 45
9 10 13   22 20 18   56 52 32

2   Add the numbers to find the total.

a $2 + 4 + 4 =$ _____   b $2 + 4 + 8 =$ _____   c $4 + 4 + 8 =$ _____

d $8 + 8 + 4 + 4 =$ _____   e $2 + 8 + 8 + 2 + 8 + 2 =$ _____

f $4 + 4 + 8 + 8 + 2 + 8 + 4 =$ _____

Challenge 3

Read each clue to find the number.

a   We are multiples of 2, 4 and 8. We are between 20 and 50. We are _____

b   I am a multiple of 2, 4 and 8. I am less than 40. I have a 3 in the 10s place. I am _____

c   I am a multiple of 2, 4 and 8. I am also a multiple of 3. I am less than 40. I am _____

Halving to find the division facts for the 4 multiplication table

Use halving to recall the division facts for the 4 multiplication table

Challenge 1

Write the number that is half of each of these numbers.

| a | 24 | b | 46 | c | 38 | d | 64 | e | 88 | f | 16 |

| g | 36 | h | 18 | i | 62 | j | 32 | k | 48 | l | 44 |

Challenge 2

Divide each array by 4. Use the halve and halve again strategy to work out each answer. Write the division calculations for each array.

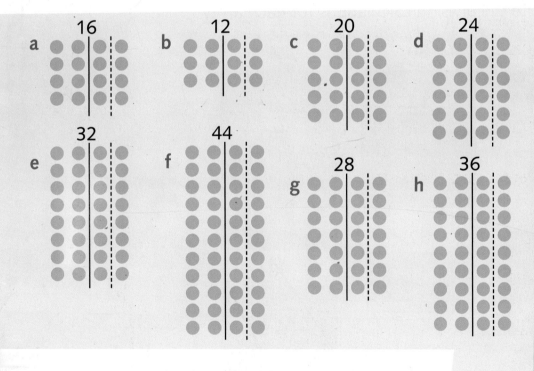

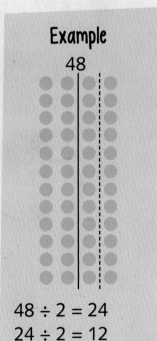

Example

48

$48 \div 2 = 24$
$24 \div 2 = 12$
So, $48 \div 4 = 12$

Challenge 3

Answer these division facts about 4 using the halve and halve again strategy. Show your working.

a $60 \div 4 =$ ☐

b $72 \div 4 =$ ☐

c $64 \div 4 =$ ☐

d $56 \div 4 =$ ☐

e $76 \div 4 =$ ☐

f $92 \div 4 =$ ☐

Halving to find the division facts for the 8 multiplication table

Use halving to recall the division facts for the 8 multiplication table

Challenge 1

Write the number that is half of these numbers, then halve the number again.

a 16   b 48   c 24   d 40   e 12

f 32   g 20   h 36   i 28   j 80

Challenge 2

Divide each array of chocolate bars by 8. Use the halve, halve again and then halve again strategy to work out each answer. Write the division calculations for each array.

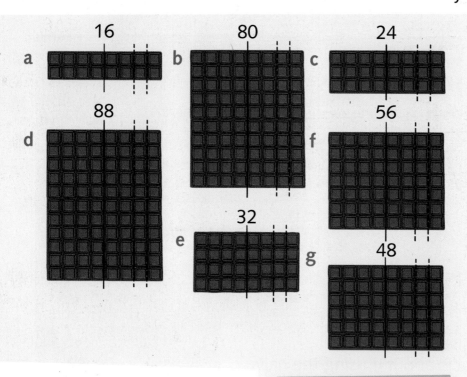

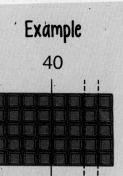

Example

40

$40 \div 2 = 20$
$20 \div 2 = 10$
$10 \div 2 = 5$
So, $40 \div 8 = 5$

Challenge 3

Answer these division facts about 8. Halve each number three times to find the answer. Show your working.

a $112 \div 8 =$ ☐   b $128 \div 8 =$ ☐

c $104 \div 8 =$ ☐   d $120 \div 8 =$ ☐

e $144 \div 8 =$ ☐   f $168 \div 8 =$ ☐

Solving word problems

Solve word problems and reason mathematically

1 Write the answers to these number facts.

a 4 x 8   b 6 x 4   c 64 ÷ 8   d 36 ÷ 3

e 9 x 4   f 72 ÷ 8   g 27 ÷ 3   h 9 x 8   i 56 ÷ 8

2 Look at the drinks for sale.
Read each question below.
Decide which operation to use.
Write the calculation, then write
the answer to the problem.

£3

£4

£8

£16

a Matt buys 7 single cups of juice.
How much does he pay?

b Jim buys three 4-cup trays of juice
and 1 single cup of juice. How many
cups has he bought altogether?

c Fran buys one 8-cup tray of juice.
How much does each cup cost?

d Mary buys two 8-cup trays of juice
and 3 single cups of juice. How
much does she pay?

e Jake wants to buy 25 cups of juice.
He has £50. Does he have enough
money? Explain your answer.

f Is it better to buy four 2-cup trays
of juice or one 8-cup tray of juice?
Why?

3 Use the pictures above to make up your own word problems with the
calculations below.

a 6 x 8 =   b 72 ÷ 8 = 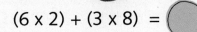   c (4 x 3) + (1 x 3) = ◯

d 7 x ● = 28   e (6 x 2) + (3 x 8) = ◯

Fractions and division

Recognise, find and write unit fractions of a set of objects

Challenge 1

Find half of each set of stamps.

a  b c d

e f g

Example

$\frac{1}{2}$ of 6 = 3

Challenge 2

1 Find half of these numbers. Write your answer as a division calculation and as a fraction calculation.

a 14   b 18   c 24   d 28   e 34

f 36   g 40   h 46   i 42   j 54

Example

12 ÷ 2 = 6
$\frac{1}{2}$ of 12 = 6

2 Find a quarter of these numbers. Write your answer as a division calculation and as a fraction calculation.

a 12   b 20   c 24   d 32   e 40   f 48

Example

12 ÷ 4 = 3
$\frac{1}{4}$ of 12 = 3

3 Find a third of these numbers. Write your answer as a division calculation and as a fraction calculation.

a 18   b 27   c 36   d 21   e 42   f 48

Example

24 ÷ 3 = 8
$\frac{1}{3}$ of 24 = 8

4 Explain how you know what to divide by to find a fraction of any number.

Challenge 3

1 What fractions of each of these numbers can you find? Write all the possibilities. Write your answers as a division calculation and as a fraction calculation.

a 15   b 20   c 24   d 27   e 32   f 40

2 Write three top tips for finding fractions of numbers.

Fraction snakes

Investigate non-unit fractions

You will need:
- coloured pencils
- squared paper

1 Look at these snakes. Write the fractions that describe each snake.

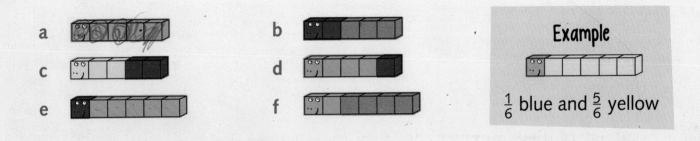

a

c

e

b

d

f

Example

$\frac{1}{6}$ blue and $\frac{5}{6}$ yellow

2 Draw your own snakes 7 squares long. Colour them in using 2 different colours. Write the fractions that describe each snake.

1 Look at these snakes. Write the fractions that describe each snake.

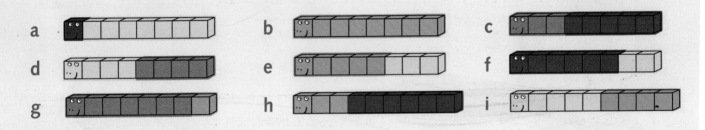

a

d

g

b

e

h

c

f

i

2 Draw snakes 10 squares long. A snake 10 squares long can be described using tenths and fifths. Colour the snakes in as many different ways as you can and write the fractions that describe each snake.

1 Take a handful of cubes in two different colours and use them to make a snake. Draw the snake. Write the fractions.

You will need:
- 20 interlocking cubes

2 Take a handful of cubes in three different colours and use them to make a snake. Draw the snake. Write the fractions.

21

Ordering fractions

Compare and order unit fractions and fractions with the same denominator

Challenge 1

Write the fraction that describes the coloured part of each circle.

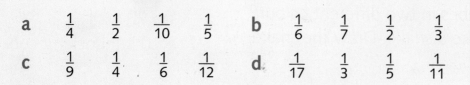

1 a   b   c   d   2 a   b   c   d

3 a   b   c   d   4 a   b   c   d

5   Now put each set of fractions in order, smallest to largest.

Challenge 2

1   Order each set of fractions, smallest to largest.

a $\frac{3}{4}$ $\frac{1}{4}$ $\frac{2}{4}$ $\frac{4}{4}$ b $\frac{3}{8}$ $\frac{6}{8}$ $\frac{1}{8}$ $\frac{2}{8}$

c $\frac{4}{9}$ $\frac{8}{9}$ $\frac{2}{9}$ $\frac{5}{9}$ d $\frac{4}{6}$ $\frac{2}{6}$ $\frac{5}{6}$ $\frac{6}{6}$

2   Write instructions for ordering fractions when the denominators are all the same.

3   Would you rather have $\frac{2}{8}$ or $\frac{5}{8}$ of a chocolate cake? Explain why.

Challenge 3

1   Order each set of fractions, smallest to largest.

a $\frac{1}{4}$ $\frac{1}{2}$ $\frac{1}{10}$ $\frac{1}{5}$ b $\frac{1}{6}$ $\frac{1}{7}$ $\frac{1}{2}$ $\frac{1}{3}$

c $\frac{1}{9}$ $\frac{1}{4}$ $\frac{1}{6}$ $\frac{1}{12}$ d $\frac{1}{17}$ $\frac{1}{3}$ $\frac{1}{5}$ $\frac{1}{11}$

2   Write instructions for ordering fractions with different denominators.

3   What does the denominator tell us about the fraction?

Fractions on number lines

Write fractions on number lines

Write the missing fractions on these number lines.

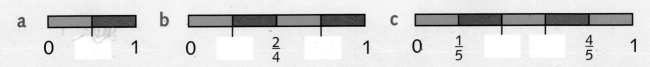

a   0   □   1   b   0   □ $\frac{2}{4}$ □   1   c   0 $\frac{1}{5}$ □   □ $\frac{4}{5}$ 1

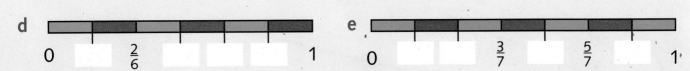

d   0   □ $\frac{2}{6}$ □   □   □   1   e   0   □   □ $\frac{3}{7}$ □ $\frac{5}{7}$ □   1

f   0 $\frac{1}{8}$ □   □ $\frac{4}{8}$ □   □   □   1

1 Complete these fraction number lines.

a   quarters   0 □ □ □ 1   b   thirds   0 □ □ 1

c   fifths   0 □ □ □ □ 1

d   eighths   0 □ □ □ □ □ □ □ 1

2 Write the next four fractions after 1.

a $\frac{3}{4}$, 1, □ , □ , □ , □   b $\frac{4}{6}$, $\frac{5}{6}$, 1, □ , □ , □ , □

c $\frac{6}{8}$, $\frac{7}{8}$, 1, □ , □ , □ , □   d $\frac{2}{3}$, 1, □ , □ , □ , □

e $\frac{8}{10}$, $\frac{9}{10}$, 1, □ , □ , □ , □   f $\frac{7}{9}$, $\frac{8}{9}$, 1, □ , □ , □ , □

Draw two number lines and write these fractions on them.

a $\frac{1}{2}$, $\frac{1}{4}$, $\frac{3}{4}$, $\frac{1}{3}$, $\frac{2}{3}$ b $\frac{1}{2}$, $\frac{1}{4}$, $\frac{3}{4}$, $\frac{1}{8}$, $\frac{3}{8}$, $\frac{5}{8}$, $\frac{7}{8}$

Measuring in centimetres

Use a ruler to draw and measure lines to the nearest centimetre

**Challenges
1, 2**

You will need:
• ruler

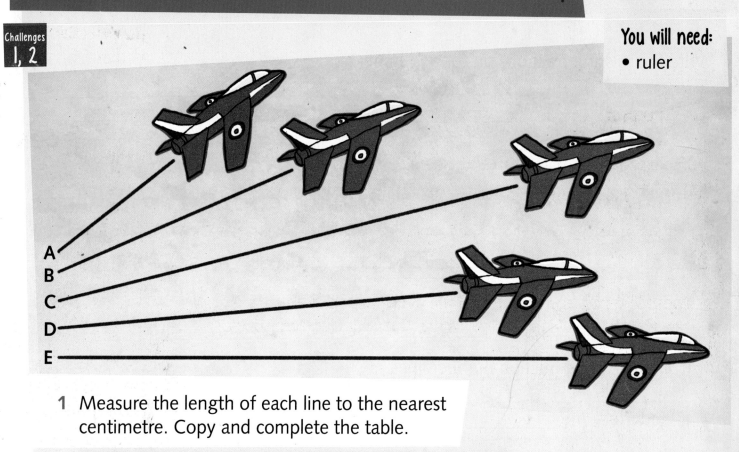

1 Measure the length of each line to the nearest centimetre. Copy and complete the table.

Line	A	B	C	D	E
Length in cm					

2 Draw lines that are 5 cm longer than lines A and B.

3 Draw lines that are $2\frac{1}{2}$ cm shorter than lines C, D and E. Write the new length above the line.

Challenge 3

Henry said, "If I take a length of ribbon from each basket and sew them together, I can make 6 different lengths."

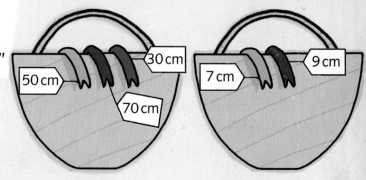

a Is he correct? Investigate.

b What is the longest length of ribbon he can make?

Measuring in millimetres

Use a ruler to draw and measure lines to the nearest millimetre

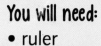

You will need:
• ruler

 1

1 Draw lines of these lengths.

a 10 mm b 30 mm c 80 mm d 55 mm e 25 mm f 75 mm

2 Measure these objects to the nearest millimetre.

a  b c

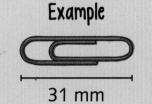

Example

31 mm

llenge 2

1 Measure each rod in millimetres, then draw a line of the same length. Below the line, write the length in two ways.

a

b

c

d

e

Example

54 mm
= 5 cm 4 mm

2 Draw lines of these lengths:

a 3 cm shorter than 85 mm   b 4 cm longer than 58 mm

llenge 3

Amy needs 4 buttons of the same size. This is her collection of buttons. Which size of button does she choose?

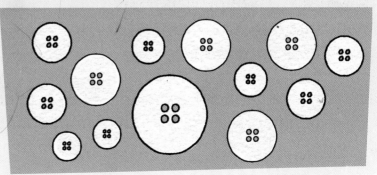

Measuring and comparing lengths

Measure and compare lengths and multiples of lengths in m and cm

 Challenge 1

Work with a partner.
Find pairs of objects.
For each pair:
- measure and
  compare their lengths
- find their total length

Example

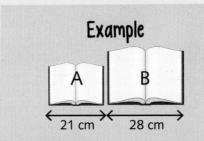

Book B is 7 cm longer
than Book A.
The total length is 49 cm.

You will need:
- ruler
- measuring tape
- collection of
  objects to
  measure

 Challenge 2

1 Work with a partner. Place the
objects on the floor. Estimate,
then measure the distance
between them in metres and
centimetres. Do this five times
and compare each measure with
its estimate.

Example

Estimate: 3 m 10 cm

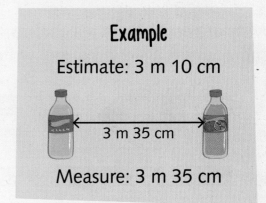

3 m 35 cm

Measure: 3 m 35 cm

2 Jim stacks boxes in a warehouse. He has a
delivery of three sizes of boxes. Find the total
height when he makes a stack of 2, 4, 5 and
10 of each box. Copy and complete the table.

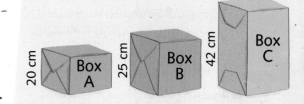

	2 boxes	4 boxes	5 boxes	10 boxes
Box A				
Box B				
Box C				

 Challenge 3

Work with a partner. Find the distance, in metres, between two objects in the
playground, for example, width of front door, length of flowerbed, distance
to gate. Your teacher will suggest suitable things to measure.

Adding and subtracting lengths

Add and subtract length using mixed units

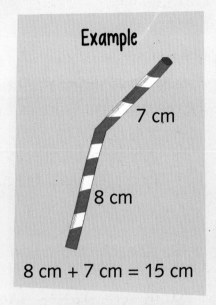

Example

7 cm

8 cm

8 cm + 7 cm = 15 cm

Challenge 1

Jim had four bendy straws, each 15 cm long. He bent each straw to make two parts. He used centimetres and millimetres to measure the parts. Copy and complete his results.

a 9 cm + ____ cm = 15 cm

b $7\frac{1}{2}$ cm + ____ cm = 15 cm

c 4 cm 5 mm + ____ cm ____ mm = 15 cm

d 12 cm 5 mm + ____ cm ____ mm = 15 cm

Challenge 2

R    S    T

1 Work out the length of Trucks R and T.

2 All three trucks board a ferry to Ireland. They park behind each other nose to tail. What is their total length?

3 A van is half the length of Truck S. What is its length?

- Truck R is 2 m 30 cm shorter than Truck S

- Truck S is 8 m 40 cm

- Truck T is $\frac{1}{2}$ m longer than Truck S

Challenge 3

There are five trucks waiting to board the ferry. The scale shows the height of each truck. Which truck is:

a 18 cm taller than Truck C?

b 25 cm short of 5 m?

c 26 cm shorter than Truck A?

d What is the difference in height between Trucks A and E?

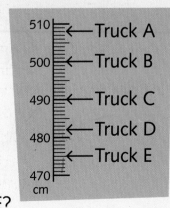

510 ←—Truck A
500 ←—Truck B
490 ←—Truck C
480 ←—Truck D
    ←—Truck E
470
cm

Expanded addition

- Add 3-digit numbers using the expanded written method of column addition
- Estimate the answer to a calculation

Challenge 1

Use the expanded method to add these numbers together. Make sure you write out each calculation correctly.

a 53 + 25	b 41 + 37	c 64 + 23	d 136 + 143
e 165 + 131	f 238 + 160	g 216 + 283	h 306 + 292

Example

```
  215
+ 342
    7
   50
  500
  557
```

Challenge 2

1 Read each calculation and quickly write an estimate of the answer.

a 285 + 417	b 258 + 431	c 392 + 307
d 445 + 313	e 427 + 436	f 308 + 395
g 417 + 387	h 226 + 479	i 516 + 247

Example

371 + 328

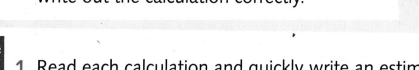
300 plus 300 is 600, and the 10s and 1s are nearly 100, so my estimate is 700.

2 Use the expanded method to work out the calculations in Question 1. Make sure you write out the calculation correctly.

Challenge 3

1 Read each calculation and quickly write an estimate of the answer.

a 406 + 389	b 517 + 378	c 605 + 389	d 468 + 474
e 547 + 463	f 738 + 255	g 179 + 808	h 643 + 397

2 Use the expanded method to work out the calculations in Question 1.

3 A pupil has written out this calculation. Explain what needs to be changed in their working out.

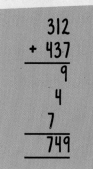
```
  312
+ 437
    9
    4
    7
  749
```

Column addition (1)

- Add 3-digit numbers using the formal written method of column addition
- Estimate the answer to a calculation

Challenge 1

1 Use the expanded method to add these numbers together.

Example
317
+ 361
8
70
600
678

a 235 + 141   b 173 + 206   c 267 + 132

d 159 + 130   e 286 + 213   f 324 + 255

2 Work out these using the formal method.

Example
317
+ 361
678

a 153 + 232   b 275 + 213   c 241 + 356

d 374 + 223   e 261 + 327   f 321 + 264

Challenge 2

1 Estimate the answers to these calculations.

a 345 + 351   b 273 + 416   c 368 + 321   d 438 + 361

e 508 + 381   f 472 + 326   g 581 + 217   h 483 + 416

2 Use the formal method to work out the calculations in Question 1.

3 Use the formal method to add these numbers.

a 326 + 347   b 259 + 435   c 415 + 319   d 536 + 248

e 447 + 344   f 539 + 353   g 263 + 518   h 366 + 326

Challenge 3

1 Estimate the answers to these calculations.

a 538 + 358   b 482 + 409   c 558 + 237   d 627 + 359

e 736 + 246   f 555 + 438   g 837 + 158   h 574 + 316

2 Explain how to estimate the answers to addition calculations.

3 Use the formal method to work out the calculations in Question 1.

Column addition (2)

- Add 3-digit numbers using the formal written method of column addition
- Estimate the answer to a calculation

Challenge 1

First estimate the answers to these calculations.
Then use the formal method to add the numbers.

a 153 + 146	b 137 + 151	c 264 + 125
d 216 + 243	e 357 + 221	f 208 + 381
g 362 + 316	h 427 + 342	i 415 + 363
j 512 + 374	k 206 + 681	l 467 + 531

Example

162
+ 237
399

Challenge 2

First estimate the answers to these calculations.
Then use the formal method to add the numbers.

Remember, always start with the 1s!

1 Carry the 1s:

a 264 + 327	b 339 + 346	c 325 + 437	d 529 + 338
e 417 + 467	f 574 + 318	g 356+529	h 608+388

2 Carry the 10s:

a 463 + 274	b 532 + 394	c 467 + 481	d 573 + 345
e 295 + 642	f 485 + 361	g 563 + 354	h 492 + 485

Challenge 3

1 First estimate the answers to these calculations.
Then use the formal method to add these numbers.

a 418 + 347	b 562 + 374	c 635 + 284	d 381 + 409
e 762 + 193	f 574 + 372	g 418 + 546	h 719 + 159

2 These calculations need the 1s and the 10s carried.

a 257 + 365	b 362 + 359	c 476 + 287	d 658 + 176

Mental addition

Add numbers mentally and use the inverse operation to check the answer

Challenge 1

Make up twelve addition calculations using these numbers.

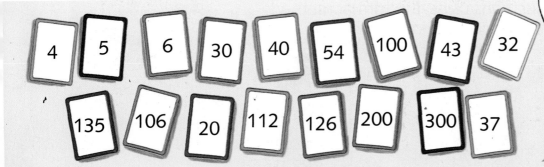

4   5   6   30   40   54   100   43   32

135   106   20   112   126   200   300   37

I like to work the calculations out mentally.

I prefer to draw an empty number line to work them out.

Challenge 2

1   Make up twelve addition calculations using these numbers.

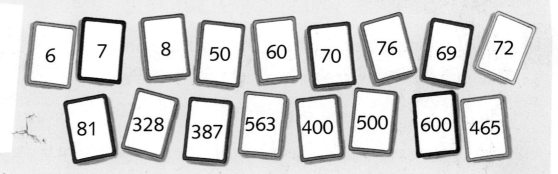

6   7   8   50   60   70   76   69   72

81   328   387   563   400   500   600   465

2   Choose four of your calculations and use subtraction to check the answers.

Challenge 3

1   Make up twelve addition calculations using these numbers.

7   8   9   94   70   60   90   79   85

88   557   684   761   845   600   700   800

2   Choose two different calculations that you can do mentally.
Explain what happens in your head when you work them out.

3   Choose four of your calculations and use subtraction to check the answers.

Column subtraction (1)

- Subtract 3-digit numbers using the formal written method of column subtraction
- Estimate the answer to a calculation

 Challenge 1

Work out these calculations using the formal method.

a 37 – 14	b 58 – 45	c 49 – 26
d 63 – 31	e 275 – 131	f 264 – 142
g 285 – 151	h 297 – 165	i 235 – 124
j 265 – 143	k 379 – 156	l 368 – 234

Example

$$\begin{array}{r} 465 \\ -\,213 \\ \hline 252 \end{array}$$

Challenge 2

1 First estimate the answers to these calculations.
Then work them out using the formal method.

a 457 – 236	b 483 – 251	c 548 – 313	d 577 – 241
e 476 – 124	f 596 – 371	g 485 – 223	h 536 – 204

2 In these calculations, you will need to change the 1s in the first number.

a 372 – 135	b 381 – 148	c 394 – 278	d 431 – 107
e 463 – 217	f 485 – 149	g 566 – 338	h 547 – 239

 Challenge 3

1 Estimate the answers to the calculations below.

2 Explain how to estimate the answers to subtraction calculations.

3 Use the formal method to work out these calculations.

a 583 – 267	b 645 – 128	c 696 – 358	d 751 – 425
e 736 – 329	f 792 – 454	g 863 – 517	h 974 – 735

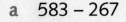

Column subtraction (2)

- Subtract 3-digit numbers using the formal written method of column subtraction
- Estimate the answer to a calculation

1 First estimate the answers to these calculations.
Then use the formal method to subtract the numbers.

a 256 – 134	b 275 – 121	c 298 – 163	
d 267 – 142	e 354 – 102	f 378 – 243	
g 356 – 124	h 384 – 251	i 468 – 147	

Example

$$478$$
$$-\ 353$$
$$\overline{125}$$

Example

$$\overset{515}{\cancel{658}}$$
$$-\ 274$$
$$\overline{384}$$

2 First estimate the answers to these calculations.
Then use the formal method to subtract the numbers.

1 Change the 1s:

a 273 – 145	b 361 – 227	c 457 – 329	d 383 – 146

2 Change the 10s:

a 326 – 142	b 365 – 183	c 348 – 271	d 437 – 253
e 516 – 372	f 538 – 276	g 657 – 382	h 614 – 283

3

1 First estimate the answers to these calculations.
Then use the formal method to subtract these numbers.

a 439 – 256	b 518 – 353	c 654 – 217	d 582 – 355
e 464 – 231	f 627 – 156	g 758 – 429	h 708 – 341

2 These calculations need the 1s and the 10s changed. Work with a partner.

a 432 – 156	b 547 – 268	c 642 – 374	d 675 – 297
e 773 – 212	f 857 – 479	g 863 – 584	h 815 – 639

Mental subtraction

Subtract numbers mentally and use the inverse operation to check the answer

Make up twelve subtraction calculations using these numbers.

I like to work the calculations out mentally.

I prefer to draw an empty number line to work them out.

65 - 7
427 - 8
361 - 40
542 - 300

4   6   6   20   30   40   47   41   54

59   167   152   148   163   100   200   300

1 Make up twelve subtraction calculations using these numbers.

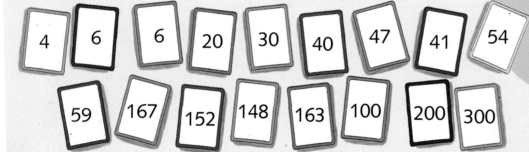

6   7   8   50   60   74   65   86   79

361   427   542   300   400   477   500   40

2 Choose four of your calculations and use addition to check the answers.

1 Make up twelve subtraction calculations using these numbers.

7   8   9   97   60   70   80   76   83

700   92   657   685   861   816   953   500   600

2 Choose two different calculations that you can do mentally.
Explain what happens in your head when you work them out.

3 Choose four of your calculations and use addition to check the answers.

Drawing and calculating perimeters

Using a ruler, draw and calculate the perimeter of rectangles

On 1 cm squared paper draw different rectangles with perimeters of:

   a 10 cm   **b** 14 cm

   c 18 cm

Example

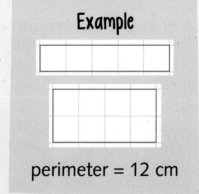

perimeter = 12 cm

You will need:
• ruler
• 1 cm squared paper

---

llenges 3

On 1 cm square dot grid paper draw a square then all the rectangles which have a perimeter of:

   a 4 cm   **b** 8 cm

   c 12 cm   **d** 16 cm

   e 20 cm

Example

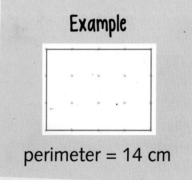

perimeter = 14 cm

You will need:
• ruler
• 1 cm square dot grid paper

---

llenge 3

Do Challenge 2 then copy and complete the table.

You will need:
• ruler
• 1 cm square dot grid paper

Side of square	Perimeter of square	Number of rectangles with same perimeter	Total number of squares and rectangles
1 cm	4 cm	0	1
2 cm	8 cm		
3 cm			
4 cm			
5 cm			

Regular perimeters

Measure and calculate the perimeter of 2-D shapes

You will need:
• ruler

Challenge 1

Use your ruler to measure the perimeter of these regular shapes in centimetres.

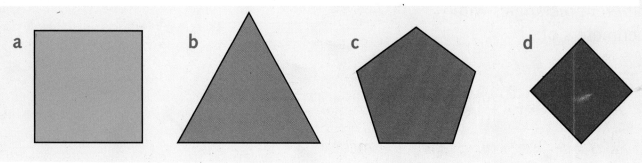

a   b   c   d

Challenge 2

Use your ruler to measure the perimeter of these regular shapes in centimetres.

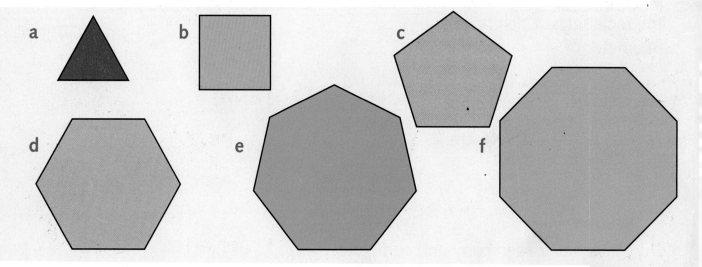

a   b   c

d   e   f

Challenge 3

1 Copy and complete the table for the regular shapes in Challenge 2.

Number of sides of shape	3	4	5	6	7	8
Perimeter in centimetres						

2 If you continue the pattern, what will the perimeter of these regular shapes measure?

a   10 sides   b   12 sides

Perimeters of 2-D shapes

Measure and calculate the perimeter of 2-D shapes

You will need:
- four to five square tiles
- 1 cm squared paper
- ruler
- coloured pencils

llenge 1

1 Make each shape with four square tiles then draw the shape on to 1 cm squared paper.

a  b  c  d

2 Below each shape, write its perimeter.

llenge 2

1 Make each shape with five square tiles then draw the shape on to 1 cm squared paper.

a  b   c   d

e   f   g

2 Below each shape, write its perimeter.

llenge 3

With five squares, you can make 12 different shapes. Seven of the shapes are drawn in Challenge 2.

1 Find the remaining five shapes.

2 Draw them on to 1 cm squared paper and calculate their perimeters.

Maths facts

Problem solving

The seven steps to solving word problems

1 Read the problem carefully.   **2** What do you have to find?

3 What facts are given?   **4** Which of the facts do you need?   **5** Make a plan.

6 Carry out your plan to obtain your answer.   **7** Check your answer.

Number and place value

100	200	300	400	500	600	700	800	900
10	20	30	40	50	60	70	80	90
1	2	3	4	5	6	7	8	9

Addition and subtraction

Number facts

+	0	1	2	3	4	5	6	7	8	9	10
0	0	1	2	3	4	5	6	7	8	9	10
1	1	2	3	4	5	6	7	8	9	10	11
2	2	3	4	5	6	7	8	9	10	11	12
3	3	4	5	6	7	8	9	10	11	12	13
4	4	5	6	7	8	9	10	11	12	13	14
5	5	6	7	8	9	10	11	12	13	14	15
6	6	7	8	9	10	11	12	13	14	15	16
7	7	8	9	10	11	12	13	14	15	16	17
8	8	9	10	11	12	13	14	15	16	17	18
9	9	10	11	12	13	14	15	16	17	18	19
10	10	11	12	13	14	15	16	17	18	19	20

+	11	12	13	14	15	16	17	18	19	20
0	11	12	13	14	15	16	17	18	19	20
1	12	13	14	15	16	17	18	19	20	
2	13	14	15	16	17	18	19	20		
3	14	15	16	17	18	19	20			
4	15	16	17	18	19	20				
5	16	17	18	19	20					
6	17	18	19	20						
7	18	19	20							
8	19	20								
9	20									

Number facts

+	0	10	20	30	40	50	60	70	80	90	100
0	0	10	20	30	40	50	60	70	80	90	100
10	10	20	30	40	50	60	70	80	90	100	110
20	20	30	40	50	60	70	80	90	100	110	120
30	30	40	50	60	70	80	90	100	110	120	130
40	40	50	60	70	80	90	100	110	120	130	140
50	50	60	70	80	90	100	110	120	130	140	150
60	60	70	80	90	100	110	120	130	140	150	160
70	70	80	90	100	110	120	130	140	150	160	170
80	80	90	100	110	120	130	140	150	160	170	180
90	90	100	110	120	130	140	150	160	170	180	190
100	100	110	120	130	140	150	160	170	180	190	200

+	110	120	130	140	150	160	170	180	190	200
0	110	120	130	140	150	160	170	180	190	200
10	120	130	140	150	160	170	180	190	200	210
20	130	140	150	160	170	180	190	200	210	220
30	140	150	160	170	180	190	200	210	220	230
40	150	160	170	180	190	200	210	220	230	240
50	160	170	180	190	200	210	220	230	240	250
60	170	180	190	200	210	220	230	240	250	260
70	180	190	200	210	220	230	240	250	260	270
80	190	200	210	220	230	240	250	260	270	280
90	200	210	220	230	240	250	260	270	280	290
100	210	220	230	240	250	260	270	280	290	300

Written methods – addition

Example: 548 + 387

Expanded written method

```
  548
+ 387
   15
  120
  800
  935
```

Formal written method

```
  548
+ 387
  935
  1 1
```

Written methods – subtraction

Example: 582 – 237

Formal written method

```
     7 12
  5 8 2
–   2 3 7
    3 4 5
```

Multiplication and division

Number facts

x	2	3	4	5	8	10
1	2	3	4	5	8	10
2	4	6	8	10	16	20
3	6	9	12	15	24	30
4	8	12	16	20	32	40
5	10	15	20	25	40	50
6	12	18	24	30	48	60
7	14	21	28	35	56	70
8	16	24	32	40	64	80
9	18	27	36	45	72	90
10	20	30	40	50	80	100
11	22	33	44	55	88	110
12	24	36	48	60	96	120

Written methods – multiplication

Example: 63×8

Partitioning

$$63 \times 8 = (60 \times 8) + (3 \times 8)$$
$$= 480 + 24$$
$$= 504$$

Grid method

×	60	3	
8	480	24	= 504

Expanded written method

```
    6 3
×     8
    2 4   ( 3 × 8)
  4 8 0   (60 × 8)
  5 0 4
  1
```

Formal written method

```
    6 3
×   ₂8
  5 0 4
```

Written methods – division

Example: $92 \div 4$

Partitioning

$$92 \div 4 = (80 \div 4) + (12 \div 4)$$
$$= 20 + 3$$
$$= 23$$

Expanded written method

```
      2 3
  4 ) 9 2
      8 0   20 × 4
      1 2
      1 2   3 × 4
        0
```

Formal written method

```
      2 3
  4 ) 9 ¹2
```

Fractions

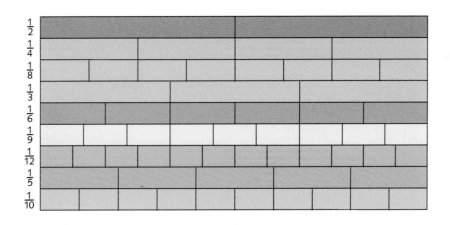

$\frac{1}{2}$	
$\frac{1}{4}$	
$\frac{1}{8}$	
$\frac{1}{3}$	
$\frac{1}{6}$	
$\frac{1}{9}$	
$\frac{1}{12}$	
$\frac{1}{5}$	
$\frac{1}{10}$	

Measurement

Length
1 metre (m) = 100 centimetres (cm) = 1000 millimetres (mm)

Mass
1 kilogram (kg) = 1000 grams (g)

Capacity
1 litre (*l*) = 1000 millilitres (ml)

Time

1 year	=	12 months
	=	365 days
	=	366 days (leap year)
1 week	=	7 days
1 day	=	24 hours
1 hour	=	60 minutes
1 minute	=	60 seconds

12-hour clock

24-hour clock

Properties of shape

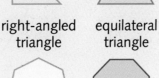

right-angled triangle   equilateral triangle   isosceles triangle   scalene triangle

circle   semi-circle   pentagon   hexagon   heptagon   octagon   square   rectangle

cube   cuboid   cone   cylinder   sphere   triangular prism   triangular-based pyramid (tetrahedron)   square-based pyramid